The New Art of War, Tactics, and Power

A New Rendition of Teachings from Sun Tzu's *The Art of War*, Niccolo Machiavelli's *The Prince*, Baltasar Gracian's *The Art of Worldly Wisdom*, and the Works of Han Fei Tzu

<u>Other Titles by Rodney Ohebsion:</u>

A Collection of Wisdom

Athletes That Inspire Us: The Remarkable True Life Stories of 20 Unique Athletes

The Get-To-The Point Success Reader 1: Selections from the Writings of Napoleon Hill, Orison Swett Marden, Samuel Smiles, Herbert N. Casson, and Charles F. Haanel

The Get-to-the-Point Success Reader 2: Selections From the Writings of Dale Carnegie, Orison Swett Marden, Charles M. Schwab, James Allen, Emile Coue, and William Walker Atkinson

<u>Coming Soon by Rodney Ohebsion:</u>

The Best of the Analects

The Forum

World Essays

Immediex Publishing

www.immediex.com

The New Art of War, Tactics, and Power

A New Rendition of Teachings from Sun Tzu's *The Art of War*, Niccolo Machiavelli's *The Prince*, Baltasar Gracian's *The Art of Worldly Wisdom*, and the Works of Han Fei Tzu

Rodney Ohebsion

Immediex Publishing

Immediex Publishing

www.immediex.com

Published in the United States

Printed in the United States

Ohebsion, Rodney

The New Art of War, Tactics, and Power: A New Rendition of Teachings from Sun Tzu's *The Art of War*, Niccolo Machiavelli's *The Prince*, Baltasar Gracian's *The Art of Worldly Wisdom*, and the works of Han Fei Tzu

ISBN 1-932968-23-7

Category: Leadership & Management

Contents

Part 3: Based on Baltasar Gracian's *The Art of Worldly Wisdom*

Part 4: Based on the works of Han Fei Tzu

Introduction

You are holding the most powerful manual of tactics, leadership, and power the world has ever known.

In this book, I have created a new rendition of material from four classic and legendary books.

First is *The Art of War*, an Ancient Chinese text based on tactics formulated by military strategist Sun Tzu about 2500 years ago. It has been prized for centuries for its valuable insights that are applicable not only to military operations, but also to most kinds of strategic endeavors.

Next is *The Prince*, a guide on how to be an effective ruler and stay in power, written by Italian political theorist Niccolo Machiavelli in 1513.

And then there is *The Art of Worldly Wisdom*, a collection of various life strategy principles written by Spanish Jesuit and philosopher Baltasar Gracian in 1647.

And last but certainly not least is the *Han Fei Tzu*, an Ancient Chinese text written by Legalist (Fa Chia) philosopher Han Fei Tzu, dealing with subjects such as authority, laws, rewards, punishments, and persuasion.

I have taken key ideas and selections from these four texts, made some new variations, additions, and condensations, and put everything in a new organization. The result is a fresh and unique version of some classic and incredibly effective teachings.

Collectively, I have titled this entire work *The New Art of War, Tactics, and Power*. Altogether, it represents some of the finest insights the world has ever produced on subjects such as strategy, leadership, management, power, and persuasion.

Part 1:

Based on Sun Tzu's *The Art of War*

Basics

The art of war is of utmost importance to the state. It will make the difference between success and failure, and between life and death. It must not be neglected.

In order to perceive field conditions, one's plans should take into account these factors that govern the art of war:

1. The Way
2. The environmental conditions and fluctuations
3. The terrain and its characteristics
4. Proper leadership
5. Proper method
6. Proper discipline
7. Proper organization, communication, and synergy
8. Proper use of paths
9. Proper finances

These factors should be familiar to every leader. Those who know them will win, those who do not will lose.

When you are deliberating conditions, ask yourself:

1. Which side is embodying the Way?
2. Which side will have advantages based on the environmental characteristics and fluctuations?
3. Which side will have advantages based on the terrain and its characteristics?
4. Of the two generals, who is most skilled?
5. Which side has more rigorously enforced discipline?
6. Which side has the stronger army?
7. Which side has better trained army?
8. Which side has constancy in conferring rewards and handing out punishments?

I can determine who will win or lose based on these factors.

Modifying Tactics

The leader who listens to me and puts my wisdom into practice will conquer, but those who do not will suffer defeat.

While listening to my ways, avail yourself to any effective circumstances and tactics that go over and beyond the ordinary rules.

Base your plans and methods according to circumstances being favorable.

How victory may be produced for them out of the enemy's own tactics—that is what the multitude cannot comprehend. Everyone can see the tactics that I use to conquer, but few can see the strategy out of which victory is evolved.

Do not repeat the exact same tactics just because they have gained you one victory—instead, let your methods be regulated by the infinite variety of circumstances.

Effective military tactics are like water. Water in its natural course runs away from high places and hastens downwards. So in war, the effective Way is to avoid what is strong and to strike at what is weak.

Just like water shapes its course according to the nature of the ground over which it flows, so also does the soldier work out his victory in relation to the foe that he is facing, the environmental conditions, and the terrain. Just like water retains no constant shape, so also are there no constant conditions in warfare.

Those who can modify their tactics in relation to these conditions, and thereby succeed in winning—I call them masterful captains.

There are a variety of circumstances, and everything is always changing. The five elements—water, fire, wood, metal, and earth—are not always equally predominant. The four seasons—spring, summer, winter, and fall—make way for each other in turn. There are short days and long. The moon has its periods of waning and waxing.

Calculations and Preparations

In military method, we have:

1. *Conditions*;
2. *Measurements*, which are based on conditions;
3. *Estimating quantities*, which are based on measurements;
4. *Calculations*, which are based on estimation of quantity;
5. *Balancing of chances*, which is based on calculations;
6. *Completing victory*, which is based on balancing of chances.

The winning general will usually make many calculations. The losing general will usually make few calculations. Calculations are an important part of preparation leading to victory. A lack of calculations can lead to defeat. So make calculations, or else you will put yourself in a major disadvantage.

Time is precious and preparation is vital; thus, the skillful soldier should avoid unnecessarily having to wait for reinforcements, or having to abandon an advantageous spot just because he needs to go back and get supplies that he should have brought with him in the first place.

Do whatever you can so that you can utilize time to put yourself ahead of your opponent. To do this you must utilize supply management.

And also, it is of utmost importance to forage on the enemy and its supplies. Eventually you must do this, and it will result in immense efficiencies compared to having to bring your own supplies, often in a time and energy savings ratio of twenty to one.

Prolonged Warfare

When you are actually in the midst of your fighting, do your best not to dissipate energy or unnecessarily put yourself in way of a long, exhaustive and draining battle—because even if you win these types of battles, they are often unprofitable in the long run. When you put yourself in such extreme conditions, this will open the possibility for your enemies to be able to take advantage of you at some point in the future.

This poses an interesting point. On one hand, everyone has heard of stupid haste in war. But on the other hand, long delays and lengthy operations are rarely ever a sign of effective warfare.

So getting back to the original point, it is clear that you should avoid long, continuous and draining warfare. No country has ever benefited from it. By really knowing the evils of such kinds of war, you can understand the proper way of waging war.

In war, let you great goal be victory—not lengthy campaigns.

Deception

All warfare is based on deception.

If you are near, make it appear to your enemy that you are far. If you are far, make it appear that you are near.

Keep yourself invisible and deceptive, and don't let the enemy anticipate or see what is coming at him. Thus, he won't know what to attack or defend, he won't see your attacks coming, and your attacks will have the most effective impact. This will also confuse him and destroy his plans and methods.

It often is the highest form of discipline to appear to the opponent that you are undisciplined. It often is the highest form of strength to conceal your strength.

By altering his arrangements and changing his plans, the skillful general keeps the enemy without definite knowledge. By shifting his camp and taking circuitous routes, he prevents the enemy from anticipating his purpose.

Take advantage of any inclinations in mood your enemy has. If he gets frustrated easily, frustrate him. If he is taking it easy, get him in a panic.

Try to make him act careless and leave openings, and then you can capitalize on them and destroy him.

Attack where he is unprepared, and appear where you are not expected.

The effective strategist should try to make as many conditions be dependent on his will, and make as few as possible be dependent on the opponent's will.

Make everything according to your advantage, starting with the most beneficial advantages first.

There is a need to manipulate the opponent in order to dominate and obliterate him. Do whatever you can to weaken him and make him vulnerable.

Those who are skilled at effectively manipulating the enemy maintain deceitful appearances based on how the enemy will act. By holding out baits, you can keep the enemy moving into bad positions, and then have soldiers set up in hiding who will capitalize when the situation presents itself.

Be prepared for any of the enemy's dominant aspects.

And if the match up does not seem favorable for you, evade it.

All these methods that I am telling you lead to victory, and must not be divulged beforehand.

Onset and Timing

The onset of troops is like the rush of a torrent that will even roll stones along in its course. The quality of decision is like the well-timed swoop of a falcon that enables it to strike and destroy its victim.

Therefore, the good fighter will be effective in his onset, and precise in his timing and decisions.

Energy can be compared to the bending of a crossbow; decisions and timing can be compared to the releasing of a trigger.

Timing is always of utmost importance in warfare. Sometimes it is necessary to exercise rapidity in order to take advantage of advantages, and sometimes you need to wait in order for proper conditions to arise that will give you an advantage, or to make suitable preparations.

Strike while the iron is hot; protect yourself at all times.

Hierarchy of Conquering

If you want to base your war method on what is practical and effective, know that it is best to take the enemy's country whole and intact. To shatter and destroy it is not as good. The same goes for capturing an army, a regiment, a detachment, or a company—it is best to capture them whole than to destroy them.

Fighting and winning in every battle is not the supreme excellence. The supreme excellence consists of breaking the enemy without even fighting.

So, it is best to pull up the roots of the enemy's plans. Next best is to prevent any synergy of his organization and forces. Next best is to attack his army in the field. Worst of all is to surround and attack walled cities. So as much as you can, avoid surrounding and attacking walled cities.

If you can do so, set up low-risk long-term tactical positioning and preparation on your enemy, and then when he steps into your baits, you can conquer him in an instant without any conventional condition warfare.

Thus, the skilled leader can stop the enemy and capture cities without even fighting, and above all, can win without lengthy field operations. With his forces intact, he maintains and gains, and without even losing one person, he completes his wins. This is the way to attack and win by clever and deceptive maneuvers. It is the preferable method.

A General's Five Main Keys to Victory

A general is a necessary part of an army, and is of the utmost importance.

As a general, there are five main keys to victory in the art of war:

1. He will win who knows when to fight and when not to fight.
2. He will win knows how to be adaptive based on his force
3. He will win who has established harmony and synergy throughout his forces.
4. He will win who, having prepared himself, waits to take the enemy unprepared.
5. He will win who has military capacity, and is not interfered with by the sovereign.

Victory lies in knowing and applying these five points.

Knowledge to Make Victory Complete

You can learn the conditions of an army through its dispositions. Hide your own dispositions, and your conditions will remain secret.

In warfare, the best strategists know themselves, know their opponents, and know the conditions.

Know your own strengths and weaknesses, and keep them hidden from the enemy. Know your enemy's strengths and weaknesses. Know the terrain and the environment.

Know the current conditions of yourself, your opponent, the terrain, and environment.

Thus, if we know that our own men are in a condition to attack, but are unaware that the enemy is not open to attack, we have gone only halfway towards victory.

If we know that the enemy is open to attack, but are unaware that our own men are not in a condition to attack, we have gone only halfway towards victory.

And if we know that the enemy is open to attack, and also know that our men are in a condition to attack, but are unaware that the nature of the ground makes fighting impracticable, we have still gone only halfway towards victory.

Therefore, the experienced soldier, once in motion, is never bewildered; once he has broken camp, he is never at a loss. Hence the saying: "If you know the enemy and know yourself, your victory will not stand in doubt; if you know Heaven and know Earth, you may make your victory complete."

Offense and Defense

The best fighters of all time are those who made themselves impenetrable to being defeated by the enemy, and then waited for an opportunity to defeat the enemy.

Securing ourselves against defeat is in our own power, whereas the opportunity of defeating the enemy is actually provided by the enemy himself.

So the best fighters are able to secure themselves from defeat, but cannot be certain of completing victory and defeating the enemy, because the enemy might not present the opportunity.

Security against defeat implies defensive tactics. The ability to defeat implies offensive tactics.

Being totally fixed on the defensive is too passive. Being totally fixed on the offensive is too aggressive.

A general who is skilled at defense is hidden in the most secret recesses of the earth. He who is skilled at offense flashes forth from the highest heights of heaven.

So, on one hand we have the opportunity to protect ourselves, and on the other, a victory that is complete.

The skillful fighter puts himself into a position that makes defeat impossible, and does not miss the moment for defeating the enemy.

Therefore, in warfare, the victorious strategist first wins, and then seeks to do battle; whereas the defeated strategist first fights, and then afterwards looks for victory.

The art of war teaches us to rely not on the likelihood of the enemy's not coming, but on our own readiness to receive him

The art of war teaches us to rely not on the chance of his not attacking, but rather on the fact that we have made our position unassailable.

An effective method of warfare is to carefully adapt ourselves to the enemy's purpose. By persistently hanging on the enemy's flank, we shall succeed in the long run in killing the commander-in-chief. This is called ability to accomplish a thing by sheer cunning. On the day that you take up your command, block the frontier passes, destroy the official tallies, and stop the passage of all emissaries.

If the enemy makes a mistake and leaves himself vulnerable, capitalize on it. Take measures to hinder your opponent by seizing what he holds dear, and subtly plan to time his arrival on the ground. Walk in the path defined by rule, and adapt yourself to the enemy until you can determine to fight the decisive battle.

At first, then, exhibit the coyness of a maiden, until the enemy gives you an opening; afterwards, emulate the rapidity of a running hare, and it will be too late for the enemy to oppose you.

A whole army may be robbed of its spirit, a commander-in-chief may be robbed of his presence of mind.

Notice the fluctuations in the moods of your enemy's army, and you can determine when to fight.

A clever general avoids an opposing army when its spirit is keen, but attacks it when it is sluggish and inclined to return—this is the art of studying moods.

Disciplined and calm, awaiting the appearance of disorder and confusion amongst the enemy—this is the art of retaining self-possession.

Closing in on victory while your enemy is far from it, waiting at ease while your enemy toils, struggles, and dissipates its energy and weakens—this is the art of husbanding one's strength.

Refraining from intercepting an enemy who has great communication, good energy, and keen spirit—this is the art of studying circumstances.

It is a rule not to advance uphill against the enemy, nor to oppose it when it is coming downhill.

Do not swallow a bait offered by the enemy.

Appear at points where the enemy must expend lots of energy and resources to defend, and rapidly march to places where you are not expected and the enemy is not ready for you.

Every enemy has vulnerable spots that can be found as long as you search for them. Even an enormous whale can be killed with a tiny harpoon. If there are people whose folly has never appeared, it is only because it has never been looked for closely enough. Attack the enemy's weak points and avoid his strong points.

When a general is unable to know how strong the enemy is and what his strengths are, and when a general allows weakness of his forces to take on strength of the enemy's forces, the result is a rout.

Emerge from the void, strike at vulnerable points and unexpected quarters, and avoid places where the enemy is very well defended.

You can ensure the success of your attacks if you only attack places that are undefended. You can ensure the safety of your defense if you only hold positions that cannot be attacked. Therefore, that general is skillful in attack whose opponent does not know what to defend; and he is skillful in defense whose opponent does not know what to attack.

Through the divine art of subtlety and secrecy, we learn to be invisible and inaudible; and hence we can hold the enemy's fate in our hands. You may advance and be absolutely irresistible if you make for the enemy's weak points; you may retire and be safe from pursuit if your movements are more rapid than those of the enemy.

Avoid fighting where you are greatly outnumbered. And always fight in a way that will not allow your enemy to be unified and well-organized. Whatever way you can use to achieve this, use it. You want to make him split up his forces to make him weakened, whereas you will only split up your forces when it is to your advantage to do so.

By discovering the enemy's dispositions and remaining invisible ourselves, we can keep our forces concentrated, while the enemy's must be divided. We can form a single united body, while the enemy must split up into fractions. Hence, there will be a whole pitted against separate parts of a whole, which means that we shall be many to the enemy's few, and we will be able to attack an inferior force with a superior one.

The spot where we intend to fight must not be made known. Thus, the enemy will have to prepare against a possible attack at several different points; and since his forces will then be distributed in many directions, the numbers we shall have to face at any given point will be proportionately fewer. For should the enemy strengthen his front, he will weaken his rear; should he strengthen his rear, he will weaken his front; should he strengthen his left, he will weaken his right; should he strengthen his right, he will weaken his left. If he sends reinforcements everywhere, he will be weak everywhere.

Numerical weakness comes from having to prepare against possible attacks; numerical strength, from compelling our adversary to make these preparations against us. Knowing the place and the time of the coming battle, we may concentrate from the greatest distances in order to fight.

Those who were called skillful leaders in early times knew how to drive a wedge between the enemy's front and rear; prevent cooperation between his large and small divisions; and hinder the good troops from rescuing the bad, and the officers from rallying their men. When the enemy's men were united, they managed to keep them in disorder. When it was to their advantage, they made a forward move; when otherwise, they stopped still.

If asked how to cope with a great host of the enemy in orderly array and on the point of marching to the attack, I should say: "Begin by seizing something which your opponent holds dear; then he will be obedient to your will."

Rapidity is the essence of war: take advantage of the enemy's unreadiness, make your way by unexpected routes, and attack unguarded spots.

Make your enemy frustrated, make him doubtful, break his will, break his spirit, drain his energy, and obliterate his confidence. Give him ways to submit and become self-defeating. Make his mind your own weapon.

Direct and Indirect Maneuvers

Understanding direct and indirect maneuvers can be used to ensure your benefit. You can use the science of weak points and strong points to give your army a tremendous force and advantage that can be compared to making your army's impact like a grindstone dashed against an egg.

In all fighting, the direct method can be used for joining battle, but indirect methods must be used to secure victory.

What do I mean by direct methods? I mean those that are straightforward and common.

What do I mean by indirect methods? I mean those that are surprising and uncommon.

Indirect tactics—efficiently applied—are inexhaustible like the heavens and earth. They are unending like the flow of rivers and streams. Like the sun and moon, they end but to begin again. Like the four seasons, they pass away to return once more.

There are no more than five musical notes, yet the combinations of these five give rise to more melodies than can ever be heard. There are no more than five primary colors—blue, yellow, red, white, and black—yet in combination they produce more hues than can ever be seen. There are no more than five cardinal flavors—sour, acrid, salt, sweet, bitter—yet combinations of them yield more flavors than can ever be tasted.

In battle, there are no more than two methods of attack—the direct and the indirect—yet these two in combination give rise to an endless series of maneuvers. The direct and the indirect lead on to each other in turn like moving in a circle—you never come to an end. Who can exhaust the possibilities of their combination?

Shuai-jan

The skillful tactician can be compared the shuai-jan. The shua-jan is a snake found in the Chung Mountains. Strike at its head, and you will be attacked by its tail; strike at its tail, and you will be attacked by its head; strike at its middle, and you will be attacked by both its head and tail.

Leadership, Management, and Utilization

The clever combatant looks to the effect of combined energy, and does not require too much from individuals; hence his ability to pick out the right men and utilize combined energy. He uses each person according to his capacity, and does not expect each person to be fit for everything.

When he utilizes combined energy, his fighting men become like rolling logs or stones—for it is the nature of a log or stone to remain motionless on level ground, and to move when on a slope; if four-cornered, to come to a standstill, but if round-shaped, to go rolling down. Thus, the energy developed by good fighting men is like the momentum of a round stone rolled down a mountain thousands of feet in height.

Carefully study the well-being of your men, and do not overtax them. Concentrate your energy and hoard your strength.

If you want to defeat your enemy, you must bring the proper emotion out of your soldiers, and let them know that they have much to gain by winning. Thus, it is crucial that you institute a system of rewards.

And when you capture spoils from your enemy, they should be distributed among your soldiers. This will really keep your soldiers encouraged, and make them experience firsthand that their effective actions are being rewarded.

Make your soldiers know that winning is a must, not an option.

If soldiers are punished before they have grown attached to you, they will not prove to be submissive; and, unless submissive, then will be practically useless. If, when the soldiers have become attached to you, punishments are not enforced, they will still be useless. Therefore, soldiers must be treated in the first instance with humanity, but kept under control by means of iron discipline. This is a certain road to victory.

Regard your soldiers as your children, and they will follow you into the deepest valleys. Look upon them as your own beloved sons, and they will stand by you even unto death. If, however, you are lenient, but unable to make your authority felt; kind-hearted, but unable to enforce your commands; and incapable, moreover, of quelling disorder; then your soldiers must be likened to spoilt children—they are useless for any practical purpose.

When the general is weak and without authority; when his orders are not clear and distinct; when there are no fixed duties assigned to officers and men, and the ranks are formed in a slovenly haphazard manner, the result is utter disorganization.

Proper communication in your maneuverings is of utmost importance. You want to move secretly, while also instituting clear organization and communication among your forces. Utilize combined energy. Any methods of proper communication should be taken advantage of.

Controlling a large force has the same principle as controlling a few men—it is merely a question of sufficiently organizing their numbers. Fighting with a large army under your command is no different from fighting with a small one—it is merely a question of having proper communication by instituting signs and signals.

It makes operations lack order when a leader commands the army to advance or to retreat, but is unknowing that it cannot obey.

It makes soldiers restless when a leader tries to govern an army the same way one would govern a kingdom. Army conditions are not the same as kingdom conditions.

It makes soldiers lack confidence when a leader is not prudent in picking the army officers.

It makes soldiers lack confidence when a team does not adhere to the principle of adapting to circumstances.

There are five dangerous faults that may affect a general:

Recklessness, which leads to destruction;

Cowardice, which leads to capture;

A hasty temper, which can be provoked by insults;

A delicacy of honor, which is sensitive to shame; and

Over-solicitude for his men, which exposes him to worry and trouble.

These are the five besetting sins of a general, ruinous to the conduct of war. Let them be a subject of meditation.

The enlightened ruler is heedful; the effective ruler is cautious.

Controlling Fight Schedules

When we are in a great position to fight and our enemy does not want to fight us, we can make him have to fight us. Just attack him in a place that he has no choice but to respond to, and then you have taken a step into drawing him into a full-fledged fight.

When we are not in a position to fight, we can prevent our enemy from engaging us in battle. Just throw something odd and unaccountable in his way.

If the enemy is stronger in numbers, we can prevent him from fighting. Figure out a way to discover his plans and how likely they are to work. Rouse him, and then learn the principle of his activity or inactivity. Force him to reveal himself, so that you can find his vulnerable spots. Carefully compare the opposing army with your own, so that you may know where strength is superabundant and where it is deficient.

Maneuvering

After collecting an army and concentrating his forces, the general must blend and harmonize the different elements thereof before pitching his camp. After that, comes tactical maneuvering.

The difficulty of tactical maneuvering consists in turning the devious into the direct, and misfortune into gain. Thus, to take a long and circuitous route after enticing the enemy out of the way, and though starting after him, to contrive to reach the goal before him, shows knowledge of the trick of deviation.

Maneuvering with an army is advantageous; with an undisciplined multitude, most dangerous.

Do not move unless you see an advantage. Do not use your troops unless there is something to be gained. Do not fight unless the position is critical.

No ruler should put troops into the field merely to gratify his own self; no general should fight a battle simply out of pride. If it is to your advantage, make a forward move; if not, stay where you are.

There are roads that must be followed, times when armies must not be attacked, towns that must not be surrounded, positions that must not be contested, and commands of the sovereign that must not be obeyed.

\---

Be quick and compact like the wind.
When you are raiding and plundering, be like fire.
When you need firmness, be like a mountain.
Make your plans dark and impenetrable like the night.
Make your moves fall like a thunderbolt of lighting.
But also note that although quickness is essential, so also is pondering and deliberating your moves when necessary.

\---

Never stop in a vulnerable position. Do not stay in dangerous isolated positions. Make effective use of allies whenever opportunities present themselves.

Your decisions on when to divide or concentrate troops should be based on the circumstances.

Observe signs of the enemy, and do not be deceived. Do not move upstream to take on an enemy. Observe dust and birds moving in order to gain clues on the moves of the enemy.

If you see that the enemy has obvious advantages to be gained but does not make an effort to secure them, this almost always means they are exhausted and vulnerable.

When you capture soldiers of your enemy, figure out a way to utilize them.

Do not press a desperate foe too severely, because when a person is desperate and cornered, he will resort to almost anything.

Nine Situations

1. On your own territory, do not fight.
2. On hostile territory but not deep into it, do not stop.
3. On ground that is advantageous to both sides, do not attack.
4. On open ground where both sides can move around freely, do not try to block the enemy's way.
5. On key intersecting-path locations where several territories are connected, form alliances, because whoever occupies these areas first will have most of the Empire at his command.
6. On ground in the heart of a hostile country with reinforced cities in its rear, gather in plunder.
7. On ground where there is terrain difficult to travel through (such as mountains, forests, rugged steeps, marshes and fens), keep steadily on the march.
8. On hemmed-in ground that you entered through narrow gorges and can only exit through difficult paths that would make your troops vulnerable, be prepared to deceive and surprise.
9. On desperate ground where we can only be saved from destruction by fighting without delay, fight.

Advantages and Follow Ups

Unhappy is the fate of one who tries to win his battles and succeed in his attacks without cultivating the results of his initiative; for that causes a waste of time, and general stagnation. Hence the saying: The enlightened ruler lays his plans well ahead, and the effective general cultivates his resources.

Strategically structure resources based on what is most advantageous.

Terrain

The common report of people is often not indicative of what is truly effective, so don't get too caught up in such labels. Just be concerned with what is really effective.

The natural formation of the country can be your best ally. Know how to assess difficulties and danger.

We cannot enter into alliances with neighboring princes until we are acquainted with their designs. We are not fit to lead an army on the march unless we are familiar with the face of the country—its mountains and forests, its pitfalls and cliffs, its marshes and swamps. We shall be unable to turn natural advantages to account unless we make use of local guides.

Spies

Hostile armies may face each other for years, striving for the victory that is decided in a single day. This being so, to remain in ignorance of the enemy's condition simply because one grudges the expenditure of a hundred ounces of silver in honors and salaries, is the height of inhumanity. One who acts like that is no leader of men, no present help to his sovereign, no master of victory.

Thus, what enables the wise sovereign and the good general to strike and conquer, and achieve things beyond the reach of ordinary men, is foreknowledge. Now, this foreknowledge cannot be elicited from spirits, it cannot be obtained inductively from experience, and it cannot be gotten by any deductive calculation. Knowledge of the enemy's dispositions can only be obtained from other men. Hence the use of spies.

There are spies that are local inhabitants of a district. There are spies that are officials of the enemies. There are spies that used to be on the enemy's side but have been converted to our side.

The enemy's spies who have come to spy on us must be sought out, tempted with bribes, led away, and comfortably housed. Thus, they will become converted spies and available for our service.

Spies can gather information for us, or cause disorder to spread among our enemies.

The utmost care and reward should be granted to spies. They must be managed properly. Be subtle! Be subtle! And use your spies for every kind of business.

Part 2:

Based on Niccolo Machiavelli's *The Prince*

Human Nature

The prince should base policies on people's true nature. People are both unique and similar. There are traits commonly common of human nature, and there are unique traits to each person. Do not overgeneralize anything about anyone. Do not make any assumptions about anyone.

Now, although each person is different, and each person is a multitude within himself that differs and varies, if you asked me to make observations commonly general to human nature, I would list a few things:

In general, people are primarily interested themselves.

People are usually motivated most by their property and honor.

People are usually liars.

People often do not practice what they preach or praise.

People are usually heavily influenced by public opinion, are usually herd-minded, and are rarely ever willing to form ideas themselves and express them.

People are often prone to trusting appearances and judging from them.

People can have their loyalty won or lost.

People usually follow the money. However, relationships based on character are far more dependable than those based on money.

Foresight

Results are a matter of both what is in a person's control, and what is not in a person's control.

The wise prince does whatever he can to exercise those things that are inside of his control, and uses what is in his control to be prepared for variations outside of his control.

Through foresight and preparation, he keeps himself in beneficial positions, he guards himself against harm, and he keeps himself ready to utilize advantages when the situation arises. He minimizes the effects of factors outside of his control, and maximizes the effects of factors in his control.

A flooding river that overflows the plains can sweep away trees and buildings, move the soil from place to place, have everything fly before it, and have everything yield to its violence without being able in any way to withstand it. But, when the weather is calm, people can make provisions with defenses and barriers in such a way that if the waters rise, they may pass away by canal, and their force will not be as out of control or as dangerous. The water will only attack spots that are not prepared to resist it.

Similarly, the person who exercises foresight will make provisions for his benefit.

Suit Action to Fit the Times

Have you ever heard a saying that if you keep on doing the same thing, you will keep on getting the same result? Well, I disagree with that.

A person can be prosperous today and ruinous the next day without any change in mood, character, and action.

And people are often observed using different methods to reach possibilities they have before them such as glory and riches: one reaches such goals with caution, another with haste; one by force, another by skill; one by patience, another by impatience, and each one succeeds in reaching the goal by a different method.

And you can also observe instances where in a group of two cautious people, one attains his goal and the other fails.

So, two people working differently can bring about the same effect, and of two working similarly, one can attain his goal while the other does not.

Aside from chance and individuality, the main reason for this is timing—whether or not people conform their actions to the spirit of the times.

I believe that he who suits his action to fit the times will prosper, but he whose actions do not accord with the times will not be successful.

A person can use a method that succeeds under one set of times and circumstances, but then if the times and affairs change, he will be ruined if he does not change his course of action to something that suits the times.

And yet, people are rarely prudent enough to know how to adapt themselves to change, usually because having always prospered by acting one way, they cannot be persuaded that it is a good idea to leave it.

And so, people often do not know how to change their behavior when it is time to do so. A person who does not change his conduct with the times will run the risk of encountering problems.

Conditions change, and people are often steadfast in their ways. As long as people and conditions are in agreement, people are successful. But when they fall out, people are unsuccessful.

Lion and Fox

A prince should choose to emulate the effective points of the fox and the lion.

A lion is strong, ferocious, intimidating, and powerful.

A fox is cunning, sly, perceptive, and understanding.

A lion cannot defend himself against traps.

A fox cannot defend himself against wolves.

Therefore, it is necessary to be a fox to discover the traps, and to be a lion to terrify the wolves. Those who rely only on the effective points of either a lion or a fox are vulnerable.

Loved or Feared

So then, let's ask this question: Is it better to be loved than feared, or feared than loved? Of course, you can say that it is best to be both. But, since they rarely come together for one person, I believe that a prince who must choose between the two will find it safer to be feared than to be loved.

Love is preserved by an obligatory link in which people may break and usually do break whenever it is advantageous for them to do so. But fear, on the other hand, is preserved by the dread of punishment, which never fails.

People love according to their own will, and fear according to the will of the person they fear. And thus, the prince will find it safest to establish himself on what is in his own control, and not in the control of others.

So now we have established that being feared but not loved is preferable to being loved but not feared. However, it is best for the prince to inspire fear in such a way that, if he indeed does not win love, he at least avoids hatred. The prince can endure very well being feared while not hated; but if he is hated, that can also bring many problems.

So then, how does the prince avoid hatred? First of all, he should not interfere with people's property. After all, people forget the death of their father before they forget the loss of their inheritance.

Next, he should respect people's traditions.

Also, he should direct people's energies into private pursuits, and promote material prosperity.

Additionally, the prince should take measures in regulating his impression on people. He should not appear to be greedy, bland, lacking in character, or excessively indecisive. The prince should do what he needs to do, but show off characteristics that are commonly praised, and hide characteristics that are considered detestable. He does not necessarily need to actually be like those good characteristics he shows off; he just needs to appear to be them.

Is it difficult to deceive people in this way? No. Due to principles of human nature, anyone who seeks to deceive will always find someone who will allow himself to be deceived.

Council

The wise price should hire quality people that will be dependent on and look out for his interests. Having the right people will also make the prince look wise to his subjects.

The prince should study his people. If he sees them thinking more of their own interests than his, and seeking inwardly their own profit in everything, then he should know that these kinds of people are not suitable or trustworthy.

The prince should want to have people who are thinking in regards to the prince's benefit, and do not pay attention to matters that are not in the prince's concern.

To keep his people honest and encouraged, the prince should study them, honor them, enrich them, and share honors and cares with them—but also make them see that they cannot stand alone, that they should not go elsewhere, and that if they stick with the prince they can get more riches and honors.

If the prince and his people are this way, they can trust each other; but when it is otherwise, the relationship will be problematic.

Information

There are three kinds of intelligence: one understands things for itself, the second appreciates what others can understand, and the third understands neither for itself nor through others. This first kind is excellent, the second is good, and the third is useless.

Now comes the subject of gathering information from people and taking advice. How can a prince ensure that he does not become prone to being surrounded by flatterers? He can do this by letting people understand that telling him the truth doesn't offend him. However, when everyone is allowed to tell the prince the truth, then the respect for him lessens.

Therefore, a prudent prince should hold a third course by choosing a selected group of capable and trusted people as advisors, and giving only them the freedom of speaking the truth to him, and only on those matters that he inquires of, and of none others. He should question them upon everything, listen to their opinions, and then ultimately use his own brain, form his own conclusions, and make the final decisions of importance.

With these people, the prince should carry himself in a way that will let each one of them understand that the more freely he speaks, the more he will be preferred. But outside of these people, the prince should listen to no one, pursue what is resolved on, and be firm in his resolutions. If he acts in any other way, he will either be overthrown by flatterers, or be so frequently changed by varying opinions that he will fall into being disrespected by the people.

So, the prince should seek guidance, but only when he wishes and not when others wish. He should discourage every person from offering advice unless he asks for it. But, he should be a constant inquirer, and afterwards a patient listener concerning the things he asks about. And additionally, if he finds out that anybody, on any consideration, has not told him the truth, he should let his anger be felt.

Part 3:

Based on Baltasar Gracian's *The Art of Worldly Wisdom*

Utilizing Social Interaction and People

Have relationships with people who can teach you, and let this social interaction be a school of knowledge. This will make your companions your teachers, and combine the pleasures of social interaction with the advantages of instruction.

Everyone has their strengths and weaknesses, their good points and bad points. There is no one who cannot teach somebody something, and there is no one so excellent that he cannot be excelled.

To know how to make use of everyone is useful knowledge.

Isn't it wise to appreciate everyone, and see the good in them?

And isn't it foolish to recognize only the good or only the bad in others?

Keep helpful wits around you, for it is a great advantage to surround yourself with champions of intellect, and use other people's skills.

There are two sets of people who can guard themselves from injury: those who have learned by experiencing it at their own cost, and those who have observed it at the cost of others.

Variate

Vary your mode of action. If you have a rival, do not always do things the same way. Thus, you can distract attention.

Do not always act on first impulse, because people will soon recognize the uniformity and, by anticipating, they will frustrate your designs. Nor should you always act on second thoughts, for people will recognize the plan the second time.

It is easy to kill a bird on the wing that flies straight, and not easy to kill one that twists and turns.

The enemy is on the watch, and great skill is required to outwit him. The tactician never plays the card the opponent expects, still less the one he wants.

Know Motives

Find out each person's thumbscrew—what he likes, and what motivates him to do things. It is the means of setting his will in action.

You must know where to get at anyone. Every choice has a special motive that varies according to taste. All people idolize something—for some it is fame, for others self-interest, for most it is pleasure. Skill consists in knowing these idols in order to bring them into play.

Know a person's mainspring of motive, and you will possess the key to his will. Utilize his primary motives, which are not always the highest, but more often the lowest part of his nature.

Observe and examine a person's ruling passion, appeal to it with words, set it in motion by temptation, and you will always checkmate his freedom of will.

So Many People, So Many Tastes

One half of the world laughs at the other, and fools are they all. Everything is good or everything is bad according to who you ask. What one pursues, another persecutes. He is an insufferable ass who would regulate everything according to his ideas. Excellences do not depend on a single person's pleasure. So many people, so many tastes, all different. There is no defect that is not affected by some.

We need not lose heart if something does not please someone, for others will appreciate it; nor should their applause turn our head, for there will surely be others to condemn it. You should aim to be independent of any one opinion, of any one fashion, of any one century.

Evading and Declining

Know when and how withdraw and decline yourself to both affairs and persons—for there are unnecessary occupations that eat away precious time, and to be occupied in what does not concern you is worse than doing nothing.

It is not enough for a careful person to not interfere with others; he must also see that they do not interfere with him. One is not required to belong to others so much that he does not belong at all to oneself.

Know how to say *no*. One should not give way in everything or to everybody. Thus, knowing how to refuse is as important as knowing how to consent. This is especially the case with people of power.

Everything depends on how you do it. Some people's *no* is thought more of than the *yes* of others; for a gilded *no* is more satisfactory than a dry *yes*.

There are some people who always have *no* on their lips, whereby they make everything distasteful. *No* always comes first with them, and when sometimes they give way after all, it does them no good on account of the unpleasant beginning.

Yes and *no* are soon said, but give much to think over.

Know how to use evasion. That is how smart people get out of difficulties. They extricate themselves from the most intricate labyrinth by using some witty application of a bright remark. They get out of a serious contention by an airy nothing or by raising a smile. Most of the great leaders are well grounded in this art.

When you have to refuse something, often the most courteous way is to just change the subject. And sometimes, pretending that you don't understand proves to be the highest form of understanding.

Do not become responsible for all or for everyone;

otherwise, you become a slave—and the slave of all. Freedom is more precious than any gifts you may be tempted to give it up for.

Self

Know yourself—your talents and capacity, in judgment and inclination. Can you really master yourself without knowing yourself? There are mirrors for the face, but none for the mind—so let careful thought about yourself serve as a substitute. Keep your foundations secure and your head clear for everything.

Know your strongest quality. Know your preeminent gift—cultivate it, and it will assist the rest. Everyone would have excelled in something if he had known his strong point. Notice in what quality you surpass, and take charge of that. In some people, judgment excels, and in others, valor. Most do violence to their natural aptitude, and thus attain superiority in nothing. Time enlightens us too late of what was first only a flattering of the passions.

Know what is lacking in yourself. Many would have been great people if they had not had something lacking, and without which they could not rise to the height of perfection. It is remarkable that some people could be much better if they could be just a little better in something. They do not perhaps take themselves seriously enough to do justice to their great abilities.

Cunning and Openness

Alternate the cunning of the serpent with the openness of the dove, and combine the two in yourself, not as a scoundrel, but as a prodigy.

Use the truth, but not the whole truth. Nothing demands more caution than the truth. It requires as much to tell the truth as to conceal it.

On one hand, a single lie can destroy a whole reputation for integrity, as the deceit becomes regarded as treason, and the deceiver becomes regarded as a traitor.

But on the other hand, not all truths can be spoken, some for our own sake, others for the sake of others.

Use cunning, but do not abuse it. Do not act too delightful or boastful about it. Everything artificial should be concealed, most of all cunning, which is hated.

Honest people are usually easiest to deceive, for those who rarely deceive and lie are usually those who often trust and believe. Indeed, being deceived is not always due to stupidity, and is often due to sheer goodness.

Make use of folly; the wisest person plays this card at times. Sometimes the greatest wisdom lies in seeming not to be wise. You need not be unwise, but merely pretend to be unwise. Adapt yourself to the situation. It surely is not foolish to bluff being foolish when needed. But it is foolish to suffer from it.

Take Care With Information

Since deceit is common, our caution has to be redoubled. To go to work with caution is of great advantage in action, and there is no greater proof of wisdom.

However, caution should not be used in such an open manner that it shows itself. After all, caution arouses distrust, causes annoyance, awakens revenge, and gives rise to more ills than you would imagine.

Take care when you get information. The ear is the side door of truth, but is also the front door of lies. The truth is generally seen, rarely heard. She seldom comes in elemental purity, especially if it has come from afar—for there is always something added to it by the moods of those through whom she has passed. The passions tinge her, sometimes favorably, sometimes odiously. She always brings out people's disposition; therefore, receive her with caution from him that praises, and with even more caution from him that blames. Pay attention to the intention of the speaker; you should know beforehand on what footing he comes. Let reflection test for falsity and exaggeration.

Do not be the slave of first impressions. Some marry the very first account they hear, and all others must live with them as concubines. But as a lie has swift legs, the truth with them can find no lodging.

We should neither satisfy our will with the first object, nor our mind with the first proposition—for that is superficial. Many are like new casks that keep the scent of the first liquor they hold, be it good or bad. If this superficiality becomes known, it then gives opportunity for cunning mischief. The evil-minded rush to color the mind of the gullible. Always, therefore, leave room for a second hearing. Alexander always kept one ear for the other side. Wait for the second or even third edition of news.

Do not make mistakes about character. That is the worst and yet easiest error. Better be cheated in the price than in the quality of goods. In dealing with people, more than with other things, it is necessary to look within. Knowing people is different from knowing things. It is profound philosophy to study the depths of feeling and distinguish traits of character. People must be studied as deeply as books.

Test the Wind

Throw straws in the air to test the wind. Find how things will be perceived, especially from those whose reception or success is doubtful. One can thus be assured of its turning out well, and an opportunity is provided for going on in earnest or withdrawing entirely. By trying people's intentions in this way, the wise person knows on what ground he stands. This is the great rule of foresight in asking, in desiring, and in ruling.

Utilize Enemies

You can utilize your enemies to your advantage, and use their ill will to level mountains of difficulties that you one would otherwise need to encounter. Many have had their greatness made for them by their enemies.

Trivial Details

Do not stand on ceremony. To be strictly attentive to the minor and most trivial details is to be a bore, yet whole nations have this peculiarity. The clothing of folly is woven out of such things. Neither affect nor despise etiquette. He cannot be great who is great at such trivial matters.

Do Not Extend Blunders

Do not turn one blunder into two. It is quite usual to commit four blunders in order to remedy one, or to excuse one piece of irrelevancy by still another. A wise person may make one slip but never two, and that only in running, not while standing still.

Do not follow up a folly. Many make an obligation out of a blunder, and because they have entered the wrong path, they think it proves their strength of character to go on in it. Within they regret their error, while outwardly they excuse it. At the beginning of their mistake they were regarded as inattentive, in the end as foolish. Neither an unconsidered promise nor a mistaken resolution is really binding; yet some continue in their folly and prefer to be constant fools.

According to the Moment

Live according to the moment—our acts and thoughts and all must be determined by circumstances. Act when you may, because time and tide wait for no one. Do not live by certain fixed rules, nor let your will pledge to fixed conditions, for you may have to drink the water tomorrow that you cast away today.

There are some people so absurdly stubborn in error, that they expect all the circumstances of an action should bend to their own eccentric whims, and not vice versa. The wise person knows that the very Pole star of prudence is found in acting according to the current circumstances.

Part 4:

Based on the works of Han Fei Tzu

Rewards, Punishments, and Law

The wise ruler uses two handles to control his ministers: rewards and punishments.

Ministers fear reprimands and punishment, and are drawn away from the dread of them. Ministers like encouragements and rewards, and are drawn towards their abundance.

The wise ruler sets up a system of positions, promotions, rewards, and encouragements for the people who are excellent and apply and display their excellence. Thus, they will remain eager and motivated to achieve gain, and everybody will achieve their objectives and benefits

To do anything other than this will bring misfortune to the ruler.

We all know it is human nature to choose safety and fortune over danger and misfortune.

Let's say the ruler's ministers who are loyal to the ruler, are devoted to his advantages, and utilize their skills and wisdom to do excellent acts, only end up miserable, poor, and encountering problems for themselves and their family.

And let's say the ruler's ministers who are not loyal to the minister, and who constantly deceive him, and bribe others to procure their own personal unjust gains, only end up achieving glory, riches and prosperity for themselves and their families.

If this indeed is the case, why would anyone expect people to choose to leave a way of safety and fortune in order to choose a place of danger and misfortune? And why would any leader expect his people to right and not wrong?

If this indeed is the case, the officials will definitely think, "It is impossible for us to procure safety and fortune through being loyal, honest, proper, uncorrupt, well-serving to the ruler, obedient to the law, and not greedy for money and rank. Thus, why shouldn't we deceive the ruler, participate in bad acts, neglect the law, only act in our interests, and do things to please and gain favor of influential people?"

And then they will end up not caring at all about the ruler's objectives, nor will they care about his laws.

They will end up forming alliances to promote their own interests below, while the ruler is getting increasingly vulnerable above.

However, if the ruler avoids this scenario, and instead uses the right people in the right way, then they will enforce and clarify the laws, and discourage bad ministers. This will bring respect and esteem to the ruler, and peace and strength to the state. And with the laws clarified and enforced, this will ensure that reward and punishment are in place. And then the ruler will have a sage's good decision making, and not be forced to follow normal convention. He will base his decisions on what really exists, and examine through checking and seeing all angles.

If this indeed is the case, then all the people working for the ruler will know that they cannot find gain and safety through deceit, falsehood, and unjust profits, and they will definitely think, "If we do not avoid doing bad things, do not use our skills and strengths to aid the ruler, do not obey the law, do not stay uncorrupted, and if we aim to make unjust profits by aligning with the wrong people and going against the law, then our situation will be hopeless."

If this indeed is the case, then the way to safety and danger will be clarified, and no one will dare betray or deceive the ruler, or exploit the community for gain. So the ministers will have all the motives and the freedom to act loyally and benefit from it, and the people will do right, and the government will be strengthened.

\---

The wise ruler uses a system of advantages and disadvantages. He shows everyone what is right and what is wrong. He clarifies standards. He offers rewards that can be earned, and institutes punishments that would be avoided.

Rewards should be guaranteed and sufficient, so that people will pursue them. Punishments should be guaranteed and sufficient, so that people will avoid them. Laws should be consistent and steady, so that people will comprehend them.

If a ruler is constant in giving rewards, and does not grant pardons when giving punishments, and if he makes rewards honorable and punishments disgraceful, then everyone will try hard.

The wise ruler makes punishment definite and authority clear.

If he is excessively compassionate, then the law will be weak. If he lacks authority, then obedience will be weak. If he does not make definite penalties, then law will be ineffective.

The wise ruler is reliable in conferring rewards and honors, and making everyone utilize their ability. If the rewards and honors are not enough or not absolute, there will be lack of obedience. But if they are adequate, then the people will be willing to do almost anything.

If the ruler does not give properly, and instead gives aimlessly and heedlessly, then the ministers will take recklessly and chaotically, and will expect rewards they are not worthy of. Then excellent actions will not be regarded highly, and the excellent people will not be sufficiently rewarded. And then, it is only a matter of time before the state's resources begin to deplete, the people will be upset over this, and they will not utilize themselves to do what needs to be done.

Overuse rewards, and you will lose the people. Overuse penalties, and you will lose the people's submission and respect. And if the rewards are unable to encourage and penalties are unable to discourage, then the nation—even if large—will be vulnerable.

King Hui of Wey said to Pu P'i, "What do you associate with me?"

He replied, "I hear your Majesty's compassion and generosity."

Delighted to hear this, the King said, "So what do you think my endeavors will lead to?"

"Problems," Pu P'i replied.

The King curiously said, "But compassion and generosity are good—so why would they lead to problems?"

Pu P'i replied, "Compassion causes mercy. Generosity causes an enjoyment in conferring favors. If you are too merciful, you will neglect condemning the faulty, and if you are too enjoying of conferring favors, you will give out rewards before excellence appears. If the faulty go unpunished and the excellent are un-rewarded, then how can problems not arise?"

If positions can be attained through influential personages, and rewards and ranks can be procured by bribery, then this opens the way for misfortune.

If the ruler is fond of dishing out unjust punishments, does not maintain the law, is fond of and easily drawn into impractical arguments and debates, and participates in excessive wordiness, then this opens the way for misfortune.

With clarified law, there is strength. With neglected law, there is weakness.

In normal circumstances, everyone knows that water overwhelms fire. However, when there is a kettle between them, the fire will endure beneath, and the water will bubble and boil away on top.

In normal circumstances, everyone assumes that the government will stop wrong just like water overwhelms fire. However, if the official in charge of affirming the law acts like a kettle, then the law will only be apparent from the ruler's viewpoint, and there will be no effective way to stop wrong.

The way I have described is the way to maintain strength among the people, and wealth among the nation. A nation that runs this way will be strong and wealthy, even if small and not heavily populated. But conversely, a nation that runs contrary to this, even if large and populous, will not have the backing of the people. The soldiers will not care about the land, and the people will not identify themselves as citizens. No ruler—not even the sage rulers Yao and Shun—and no dynasty—not even the three dynasties Hsia, Yin, and Chou—could reign well and be strong in such conditions.

Orders must deal with human feelings—which have their fine points that reward and punishment must be based on.

Do Not Be Excessively Meddlesome

Even if a ruler is wise, he should not be excessively meddlesome. He should let things find their proper place.

Adapt

People who lack an understanding of government often preach that "Old methods should never be altered, and accepted customs should not be abandoned."

A sage, however, is not decidedly for or against change. All he is interested in is the proper and effective way of ruling. His decision to alter old methods or abandon accepted customs is only based on the criteria of whether or not these old methods and customs are effective right now.

The sage does not necessarily endeavor to use the ways the ancients used. And he never institutes rigid and fixed standards to be applied at all times. He observes and examines things based on the way they are now, and he adjusts to deal with them accordingly.

Verily, this is the only effective way. And so, the superior person does not follow a set principle in his speech, or an unalterable method in his actions.

There was a greet shoemaker from Lu, whose wife was a great silk weaver. They both decided to move to Yueh, but then someone told them, "You should rethink the move."

"Why?" asked the man.

The reply was, "You make shoes for the feet—but the people of Yueh walk barefooted. Your wife makes silk used for helmets, but the people of Yueh do not use them in their hair. Your skills will be useless in Yueh—that is why I said you should rethink the move."

You Can't Lift Your Own Self

There are three truths in this world: Even the wisest people will find some goals impossible, even the strongest people will find some objects unmovable, and even the mightiest people will find some opponents unbeatable

For example, even if there is someone who can match the wisdom of the Sage Emperor Yao, he will not accomplish spectacular things if he does not have the people's support.

And even if there is someone who is as strong as the legendary strongman Wu Huo, he still cannot lift his own self without the assistance of others.

And even if there is someone as powerful as Meng Pen or Hsia Yu, he still cannot win all the time if he neglects maintaining the law or using and dealing with people properly.

And so, the legendary strongman Wu Huo found 500 pounds to be light, but his own body to be heavy—not because his body weighed more than 500 pounds, but because his position would not assist him in lifting his own body.

And Li Chu found it easy to see across the longest distances accurately, but difficult to see his own eyelashes—not because the long distances were near and his own eyelashes were far, but because the natural structure would not allow him to view his own eyelashes.

And so, the wise ruler will not blame Wu Huo for not being able to lift himself, and he will not humiliate Li Chu for not being able to see his own eyelashes.

The wise ruler depends on utilizing advantageous circumstances, and he looks for the easiest way, in order to use a minor effort to achieve great deeds and garner a good reputation.

Times, circumstances, and opportunities change, matters can be advantageous or the opposite, and things come and go.

So now you know why Yao cannot rule effectively by himself, why Wu Huo cannot lift his own body alone, and why Meng Pen or Hsia Yu cannot be victorious just by themselves.

Maintain the law and deal with people properly, and then you can observe what people do properly.

The wise ruler observes what people do, but avoids letting people observe his own motives.

Using People

Tzu Chang was pulling a cart up a hill, but could not manage the weight, so he started singing. This attracted people around him, and they all helped him take the cart up the hill.

Suppose Tzu Chang did not have a method to attract people. Then even if he strained himself to exhaustion, he still wouldn't have been able to take the cart up the hill. He didn't strain himself because he had an effective method of using people.

If the ruler does not employ the state's capable people, or if he only appoints and dismisses based on reputations and not on examining excellence, then this opens the way for misfortune.

Meng Sun went hunting and got a young deer. He ordered Ch'in Hsi Pa to bring it back. On the way, the mother deer followed and cried, and Ch'in Hsi Pa found this so unbearable that he returned the baby deer to its mother.

Then when Meng Sun asked for the baby deer, Hsi Pa said, "I couldn't bear the mother's crying, so I returned it."

Greatly angered, Meng Sun immediately fired him.

However, just months later, Meng Sun rehired him and appointed him as his son's tutor.

When Meng Sun's driver heard about his, he curiously asked, "Why did you blame him earlier, and then call him back and appoint him as tutor to your own son?"

"Well," Meng Sun said, "he could not bear the ruin of a baby deer—so how could he possibly bear the ruin of my son?"

Persuasion

For the most part, the difficulty in persuading people is not due to not knowing the necessary information to plead one's view, nor is it due to lack of skill in argumentation that will make one's ideas clear, nor is it due to a lack of carefulness in fully utilizing one's abilities. For the most part, the difficulty in persuading people is found in reading someone else's mind-and-heart, and adapting your statements to conform to it.

Let's say the person you are trying to persuade is concerned with having a reputation for virtue, and you decide to discuss with him about making lots of money. He will end up thinking of you as rude, he will be neglectful to you, and probably tell you to get lost

On the other hand, if the person you are trying to persuade is concerned with making lots of money, and you decide to discuss with him about a reputation for virtue, he will consider you as tactless and unperceptive, and he will not consider your statements.

And if the person you are trying to persuade acts like he is concerned with a virtuous reputation but is actually secretly concerned with making lots of money, and then you discuss with him about a reputation for virtue, he will act like he is considering your statements and is open to you, but in reality he will disregard you; and if you discuss with him about making lots of money, he will outwardly disregard you, but will secretly regard your statements.

Thus, you should ensure that you pay close attention to these aspects.

If you discuss with a ruler about people of high caliber, he will think you are suggesting that he is inferior to them. If you talk about people of low caliber, he will think you are trying to make yourself look good so you can manipulate him.

If you discuss his likes, he will suppose that you want to take advantage of him. If you discuss what he hates, he will suppose you are attempting to meddle with his patience.

If you speak too straightforward and forthright to him, he will think you are deficient, and he will avoid you. If you speak too fancily and explanatory, he will think you are too conceited, and he will disregard you.

If you are too unspecific when you present your ideas, he will conclude you are a sissy who is too cowardly to express what he means. If you are too expressive and verbose, he will regard you as a crude vulgar person who wants to look down at him.

Such are the difficulties in persuasion—you must take heed of them.

The key to persuasion is in knowing how to feature the perspectives that the person you are talking to wants to promote, while you downplay the aspects that he wants to hide.

Let's say you want to persuade someone to take your suggestion to promote good relationships within the state. You should describe it in a way that emphasizes its magnificent aim, and affirm how it corresponds with his own interests.

Let's say you want to talk about things that present danger to the state. You should start off by mentioning the faults and negative points against them, and then correspond them with how they go against the person's own interests.

Praise other people who have similar actions to the person you are talking to, and esteem tasks that are in the same category that his tasks are.

As the ruler begins to trust you and regard you as honest but loyal, then you can be more honest in weighing pros and cons based on present conditions, and can show your excellence in actions, and be direct in showing the good and bad points of the government's methods.

In early times, Mi Tzu Hsia became popular with the ruler of Wei State. At the time, the laws of Wei State said that any person who secretly used the ruler's carriage would face punishment of having his feet cut off.

One day, someone went into the palace late at night and informed Mi Tzu Hsia that his mother was sick. Upon hearing this, Mi Tzu Hsia forged a fake request from the ruler in order to use his carriage, and then took it to go see his mother.

When the ruler found out about this, not only was he not offended, but he only had good things to say, and remarked, "What a filial child! Over his concern for his mother, he went to the extent of risking his feet being cut off!"

Another time, Mi Tzu Hsia was walking outdoors with the ruler, and began eating a peach. Tasting how delicious it was, he offered the remaining half to the ruler, who remarked, "Your love for me is truly genuine! You put your appetite aside, and are instead concerned with offering me tasty food!"

But later, when Mi Tzu Hsia was older and less attractive, and the ruler was not so enamored with him anymore, a charge was brought against him by the ruler, who remarked, "Don't forget, this guy once stole my carriage, and another time he offered me a peach that he already ate half of!"

Although Mi Tzu Hsia's actions were the same way as before, he was praised in the earlier instances, but charged with wrongdoing in the later instance—and this was all because the ruler's love for him had converted into disdain.

Anyone who wants to make pleads, persuade, explain or discuss anything with powerful people must first intently observe what that person loves or hates.

It like a squiggling dragon that you can tame, ride, and play with, but that has certain sharp points that can kill a person and must be avoided. In the same way, a powerful person has these sharp points, and a prospective persuader who avoids touching them has gone a long way in being a masterful persuader.

A Tale of Two Men

In early times, Duke Wu of Cheng planned to invade Hu. So he gave his daughter in marriage to the ruler of Hu, causing him to let down his guard.

Then he asked his ministers, "I am considering starting a military campaign. What countries should we invade?"

His High Officer Kuan Ch'i Ssu said, "We should invade Hu."

Greatly angered, Duke Wu had the man executed, and exclaimed, "Hu is our brother state. How can you suggest invading it?"

When the Ruler of Hu heard about what happened, he assumed that Cheng was on friendly terms with him, and did not institute preventive measures against a potential invasion from it.

Not much later, the people of Cheng made a surprise attack on Hu and conquered it.

There was a rich man who lived in Sung. One day, it rained intensely, causing his mud fence to topple. The man's son said, "If we don't rebuilt the fence immediately, robbers might come."

The man's neighbor also said the same thing to him.

That evening, the man was indeed robbed, and lost a great deal of property. From then on, the man and his family greatly regarded the son's judgment, but were suspicious of the neighbor.

In both of these cases, what the two men said came out true. The man in the more severe scenario was executed, while the other man in the lesser case incurred suspicion. It's not like either of them had trouble getting the appropriate information. It's just that they had troubles in using it the proper way.

Motives

A high official from one state loved warriors and scholars, and this aided to his state's well being. A high official from another state also loved warriors and scholars, and this led to his state's chaos. They both had the same love for warriors and scholars, but they had different motives for what they did.

If a crazy person is running in one direction and a person who is chasing him is also running in the same direction, then the direction they are running in is the same, but their motives for doing so are not he same. Thus, there is a saying: "People doing the same things might have different motives."

Eels are like snakes; silkworms are like caterpillars. People are afraid when they see snakes, and shocked when they see caterpillars. And yet, fishermen are willing to hold eels in their hands, and women are willing to pick up silkworms. So, when there is profit, people become as courageous as great warriors.

Earl Chih was about to attack Ch'ou Yu State, but noticed that the path was too dangerous to travel through. So, he had large bells made, and then presented them as a gift to the Ruler of Ch'ou Yu, who, greatly pleased, planned to clear the way so that they could be delivered. But Ch'ih Chang Wan Chi interjected and remarked, "Don't do it. That man is acting the way a small state respects big power. But now that it is a big state that is sending us such a gift, surely soldiers will follow it. Do not accept it."

But the Ruler of Ch'ou Yu did not heed this, and he later decided to accept the bells, and thus, Ch'ih Chang Wan Chi cut the naves of his carriage short enough for the narrow road, and drove away to Ch'i State.

And then seven months later, Ch'ou Yu was destroyed.

In Ch'i State, the people's expensive funerals used such thick coffin walls, that the Duke remarked to his advisor, "If they use up wood like this, there will be little left for the nation's military—and yet, despite this, they continue to have expensive funerals. How can this be stopped?"

His advisor remarked, "When people do anything, they are motivated by profit and/or reputation."

And thus, he immediately issued an order that if coffin walls exceeded certain legal limits, then the government would institute a punishment of chopping up the corpses, and blaming the mourning relatives. This would make expensive funerals accomplish neither reputation nor profit, and thus people would stop having them.

Li K'uei was Governor of the Upper Land under Marquis Wen of Wey, and he wanted every man in the region be a good shooter. He issued a decree that if any men were involved in an unsettled legal dispute, they would have a target shooting competition, and the winner would win the suit, while the loser would lose the suit.

As soon as the decree was issued, the whole region began practicing archery day and night continuously.

And then, when the region went to war with the Ch'ins, they obliterated them due to the fact that everyone was such a good archer.

There was a horse expert who could manage several horses at a time with incredible proficiency, controlling their speed, maneuvering them with precision, and controlling them in any way he pleased by using the whip and reins.

However, when a wild pig came jumping at the horse, then the horse expert lost control of the horses—not because the whip and reign were less severe, but because the horse expert's authority over the horses was surpassed by that of the wild jumping pig.

Even if a ruler is excellent, he should not make assumptions about acts. He should intently observe and examine what motivates ministers' actions.

Rulers and ministers have different interests, and thus, ministers can never be totally loyal.

Gathering and Controlling Information

A ruler said to a wise man, "Everyone knows that if you lack getting the proper information from enough perspectives, this can result in problems. However, in my experience managing the state, I consult with almost everybody, and the more I consult, the more disorder there is in the state. Why?"

The wise man said, "Well, in the case of a wise ruler asking the right sources for information pertaining to the state, one source might be able to ascertain the right information, while another might not. And thus, in such a case, a wise ruler can have various sources of honest and responsible information picked up and weighed against each other.

"However, in the current situation in your state, every source is biased in order to be in accordance with the opinion of one influential person—Mr. Chi. And so, with everything having the same bias, even though there are so many sources, they are all really based on the viewpoint of one person. Hence, no matter how many sources you consult with, are you really consulting with more than one source? And even if you consult with every source in the state, can you expect anything but disorder in the state?"

Fortune telling methods like using turtle shells, bamboo slips, and astrology are not capable of guaranteeing wins or foretelling military outcomes. It is very foolish to believe in such things.

In modern times, everyone knows about the teachings practiced by the Confucians and the Mohists.

The Confucians esteem supreme regards for Confucius, and the Mohists do the same for Mo Tzu.

Since the death of Confucius, eight current distinct Confucian sects have emerged and are being followed.

Since the death of Mo Tzu, three current distinct Mohist sects have emerged and are being followed.

And despite the fact that each sect has varying and sometimes contradicting teachings and practices, each of these sects insist that they have the true teaching of Confucius or of Mo Tzu.

It is clear that we cannot bring Confucius or Mo Tzu back to life—so who can assert which of the varying versions of their teachings today is the accurate one?

And as for Confucius and Mo Tzu, each of them was an adherent to the ways of the legendary ancient exalted Sage-Emperors Yao and Shun. Yet Confucius and Mo Tzu had teachings that differed from each other, and each of them indicated that they were practicing the real ways of Yao and Shun.

It is clear that we cannot bring Yao and Shun back to life—so who can assert whether it is the Confucians or the Mohists that have the accurate version?

Now consider this: it has been over seven hundred years since the Yin and early Chou time periods, and it has been more than two thousand years since the Yu and early Hsai time periods. Since we cannot even agree on which of the current versions of Confucian and Mohist teachings are accurate, which are only a few hundred years old, how can we even begin to ascertain the ways of Yao and Shun, who lived about three thousand years ago! It is clear that we cannot be sure of anything at all!

Clearly, people who assert that they are following the ways of ancient kings, and say that they are sure of their descriptions of the ways of Yao and Shun, are surely either fools or fakers!

A wise ruler will never strictly adhere to teachings that come from fools and fakers, and are so varied and contradictory.

The wise ruler looks at all sides of a situation. Otherwise, he cannot obtain what is authentic.

The wise ruler listens, but makes speakers responsible for what they say. Thus he can ascertain needed information, but does not allow ministers to neglect their abilities.

The wise ruler investigates into matters by maneuvering information. Thus he can use the known to find out the unknown.

The wise ruler maneuvers statements and matters. This allows him to make observations and find out wrongdoings.

If the ruler takes advice only from high-ranking ministers, never listens to a variety of sources, and only has one exclusive source of information, then ruin is possible.

The wise ruler, in regards to people and gossip, observes and verifies who is skilled at what and who is lacking at what, and he does not allow ministers to spread gossip and rumors.

If the ruler is foolish and easily swayed, if he has no discretion and lets every secret out, if he tells every minister what the others said, then ruin is possible.

Know, but avoid being known.

Dealing with Women, Associates, and Relatives

The wise ruler, in conducting himself with women, enjoys them, but doesn't let them manipulate and control him with their requests.

The wise ruler, in dealing with his closer associates, enjoys them, but keeps them responsible for what they say and do, and does not let them express opinions that are inappropriate and not asked for.

The wise ruler, in regards to his relatives, keeps them responsible for the results of their advice, and will not promote them erratically.

Judge Based On Realness

T'an T'ai appeared to be a superior person. Confucius regarded him as having great potential, so he took him in as a disciple. However, after interacting with him for a while, Confucius discovered that his conduct did not correspond with what he appeared to be.

Ts'ai Yu's speech was brilliant and cultivated. Confucius regarded him as having great potential, so he took him in as a disciple. However, after interacting with him for a while, Confucius discovered that his wisdom did not correspond with his speaking skill.

Thus, Confucius said, "Should I pick people based on their appearance? I made a mistake in regards to T'an-t'ai. Should I pick people based on their speech? I made a mistake in regards to Ts'ai Yu."

So even Confucius—who was supremely wise—had to acknowledge that his judgment was mistaken. And the speakers of today are more articulate than Ts'ai Yu, and the rulers of today are more easily deceived than Confucius. So if they assign people to office solely based on their satisfaction with their speech, then how are they going to avoid making mistakes?

Mistakes are easily made when people trust others just based on their appearance or their skill in speaking.

If someone only observed how much tin is in a certain mixture and what color the metal is, and did not examine it in any other way, then even the masterful Ou Yeh could not be certain of how sharp a sword is. Yet if someone observes it slice off water-bird heads and cut up land-horses, then even the most ignorant slave would be able to know that the sword is sharp.

If someone only examined the shape of a horse's teeth, then even the famed judger of horses Po Lo could not be certain of the horse's quality. Yet if someone attaches it to a carriage and observes the way it moves over a certain distance, then even the most ignorant slave would be able to know if the horse is effective.

And if someone only looked at a person's features, clothing, and speech; then even the supremely wise Confucius would not be able to say what sort of a person he is. Yet if one sees how he acts in a position and observes his actions, then even someone with so-so judgment would be able to know if he is wise or not.

Don't change the statements you hear—just compare them with the actions, and observe whether the statements and actions correspond with each other.

The superior person takes the inner feelings, but leaves the outer appearances. He is fond of the inner qualities, but hates the outer decorations.

A man wanted to purchase a new pair of shoes for himself. He measured his feet at home and then went to the marketplace, but noticed that he had forgotten to bring the measurements. He decided to go back and get them, and when he finally got back, the marketplace was closed.

He told someone what had happened, and the person said, "Why didn't you just try the shoes with your own feet?"

The man replied, "I have confidence in my measurements, but not in my own feet."

Authority Versus Righteousness

Most people will submit to authority; very few will be moved by righteousness. Consider the example of Confucius, who was one of the supreme sages in world history. He had exemplary actions and he illustrated the Way—yet as he traveled about through many areas, he only attracted 70 main disciples.

It is very uncommon to see reverence for benevolence[1] and loyalty to righteousness, and it is rather difficult for one to act thus. So in all the wide areas that Confucius traveled, he gathered only 70 main disciples; and only one person—Confucius himself—was really righteous and benevolent.

Now consider the example of Duke Ai of Lu. He was a so-so ruler, but when he rose to power as the head of the state, there was nobody throughout the territory who was unobedient to him.

People will by nature submit to authority. Anyone who seizes authority can easily make people submit. This is why Confucius stayed a citizen, and Duke Ai stayed as his ruler. It's not like Confucius was prompted by the righteousness of Duke Ai. It was simply that Duke Ai exercised authority, and thus he caused Confucius acknowledge his preeminence.

In modern times, it is common for scholars that are advising a ruler, to neglect recommending him to use authority, even though it is a sure way to effectiveness. Instead, they are adamant in telling him that he should practice benevolence and righteousness in order to be a real ruler. This is like asking him to be like Confucius, and expecting most people to become like Confucius's disciples. Having an approach like this will most likely lead to poor results.

---

[1] or self-purity / virtue

Putting Oneself in the Right Position

If many people are in a position to benefit from a ruler's misfortune, then the ruler is in a dangerous position.

A Sage-Ruler institutes a policy where the people have no way to do him wrong, and cannot avoid doing him good. He never relies on them doing him good only by love. It is dangerous to rely on the people doing him good with love. It is safe to rely on their not being able to avoid doing him good.

If he depends on people doing him good, then even if you search throughout the state boundaries, you will not even find a few dozen of such people. But if he makes sure there is no way they can do him wrong, then an entire state can be made constant.

A ruler should think more about the many and less about the rare few, so he should focus on law instead of virtue.

It is dangerous for a ruler to trust others. Anyone who trusts others can be manipulated by others.

The wise ruler, in dealing with what he wants, does not let anyone be in control of those things, because that will open the possibility for people to manipulate him.

Advantages

A ruler must take advantage of advantages, and counteract disadvantages.

Overvaluing minor advantages will impede major advantages.

If the ruler knows of a major advantage but does not pursue it, if he knows of some major misfortune in its early stages but does not do anything to stop it, and he is thus lacking adequate offense and defense, and he tries to build a foundation based solely on benevolence[1] and uprightness, then this opens the way for misfortune.

[1] or self-purity / virtue

Efficiency

If there is a firefighting team, and the captain himself is the one who transports containers of water and goes to the fire, then he is only doing one person's function, and not much will get done. But if he uses authority to give orders to the others, he can reside over many people.

So, sages and wise rulers do not neglect the big picture in order to pursue trifling things.

By pursuing the right course of an endeavor, it can be accomplished without difficulties and great efforts.

Loyalty and Confidence

Li K'uei was constantly telling his guards, "Be careful and aware, since an enemy attack might come at any time." After months of this, the enemies still had not attacked, and the guards became exhausted, did not do their duty, and also had less loyalty or confidence in Li K'uei. Not much later, the Ch'ins attacked and destroyed them. This is the misfortune of lack of loyalty or confidence.

Petty loyalty can be the betrayer of major loyalty.

If people lack confidence in the ruler, and inferiors are un-obedient towards superiors, and the ruler is too reliant on the premier, then ruin is possible.

Controlling Resources

The wise ruler, in doing favors, is in sole charge of resources, and does not allow his ministers to give them based on their own preferences.

Traits to Avoid

If the ruler is unobjective, easily provoked, imprudent, excessively sensitive, angered by all things, quick to use weapons, or if he thoughtlessly gets into wars or makes invasions, and does not properly train the military or the agricultural industry, then ruin is possible.

Chung Yung

The superior person avoids excesses and avoids deficiencies.

Avoid extremes and excesses, so you will not expose yourself to danger.

Also by Rodney Ohebsion:

A Collection of Wisdom

Over 600 pages of the world's greatest wisdom and teachings from people, quotes, classical texts, philosophies, cultures, folktales, and proverbs—all put in an unparalleled level of clarity, efficiency, and accessibility

Athletes That Inspire Us

The Remarkable True Life Stories of Twenty Unique Athletes

The Get-To-The-Point Success Reader 1

Selections from the writings of Napoleon Hill, Orison Swett Marden, Samuel Smiles, Herbert N. Casson, and Charles F. Haanel

The Get-To-The-Point Success Reader 2

Selections From the Writings of Dale Carnegie, Orison Swett Marden, Charles M. Schwab, James Allen, Emile Coue, and William Walker Atkinson

Coming Soon by Rodney Ohebsion:

The Best of the Analects

The Forum

World Essays

For information on all of these books, visit:

www.immediex.com

Printed in the United States
144962LV00005B/15/A

9 781932 968231